A
Simple
Unified Theory

A
Simple
Unified Theory
FROM MAGNETISM TO GRAVITY

H. C. Huang

Port Townsend, WA
e10hchuang@yahoo.com

Formerly entitled *Attraction or Rejection?: Postulations from Magnetism to Gravity*.

Library of Congress Control Number: 2006905609
ISBN: Hardcover   978-1-4257-6282-7
               Softcover   978-1-4257-6276-6

This book was printed in the United States of America.

To order additional copies of this book, contact:
Xlibris Corporation
1-888-795-4274
www.Xlibris.com
Orders@Xlibris.com
34558

A Story

The universe is without limit, as are our thoughts . . .

Like many others who think freely, I have silently pondered the most basic questions of our physical universe. In the pages that follow, I give expression to these thoughts. Dear scientists, I beg your forgiveness for this intrusion into your realm of knowledge. My intent is not to invade, but rather to inform. Please accept my appreciation of your pioneering spirit and respect for your earnest efforts to probe the most difficult questions at the boundaries of science. May my thinking illuminate your efforts into an exquisite future.

I have documented my theories of physics at the urging of Leadia Koo, who, for so many years, has been the only one truly interested in and patient enough to listen.

Finally, dear Leadia, my book is here.
"Salud," dear Leadia, my lady.

With my deepest gratitude,

H. C. HUANG

CONTENTS

FOREWORD

SCIENTIFIC DISCOVERIES OFFER human civilization a solid footing upon which to climb, with a greater sense of certainty, the stairway into the future. Like others before me who have probed the nature of our universe, I too have raised basic questions, whose answers—my postulations on physical phenomena—are presented in this report.

The report is divided into four interrelated parts. Part I, "Magnetic Bars: Origins of Magnetic Force," deals with the origin of the "mystic" force and its behavior, including the mechanism of repulsion and attraction. It also explains certain phenomena and changes in Earth's polarities. Next, Part II, "Photons: Wave-Particle Parallelism," considers the dual nature of light. Current scientific thinking widely accepts that photons have both a wave and particle nature. But along with Einstein, who insisted that photons are particles, I concur, proposing the foundation for this agreement. Thus, Part II discusses the wave behavior of light particles. It also mentions how light bends when it passes a gravity field, owing to the structure of the gravity field (discussed in Part IV). In Part III, "On Color Force and Color Charges," I approach the structure of the atom to derive both the strong and weak forces. Indeed, the same positive, negative, and neutral charges extend from the level of magnetic force to those of photons and atoms. Gravity is the synthesized mass of these same forces. Thus, Part IV, "Gravity's Source," discusses the structure of gravity. Since a beam of light is composed of non-continuous particles (even though they can be continued), beams of light cannot be woven into fabric to hold any mass. On the other hand, gravity holds every mass. I suggest that there are no gravitons because exchanging particles will consume energy. I also suggest how masses accelerate differently in free falls, how inertia works, and how an object maintains its linear motion.

Finally, I propose that all matter, including the photon, neutrinos, and gamma rays, has its respective mass. Perhaps the formation of differently-featured protons and neutrons occurs, but is not stable enough to persist.

—H. C. Huang
St. Patrick's Day (March 17), 2006
British Columbia

PART I

Magnetic Bars:
Origins of Magnetic Force

Section 1.
Revolving Electrons

W HEN THE FERROUS molecules inside a magnet are in straight array, the electrons surrounding some atoms in the molecules are free to revolve in the same direction, either clockwise or counterclockwise. When all revolving electrons in their respective arrays revolve uniformly in one direction, the polarities are apparent. The more orderly the arrays, the stronger the magnetic force.

Section 2.
Repulsion

When the like polarities (magnetic "S" ends) of two magnets are placed in close proximity, the electrons' rotation generates a repulsive force (Fig. a). The axis of the rotating electron is the magnetic line, or string, of force. Two spinning electrons revolving in opposite directions at their tangential contact points or phases will repel and bounce away from each other (Figs. aa 1, 2, and 3). However, it must be pointed out that, when they merge, their overall directions of movement are the same before they touch and, when they repel, their overall directions of movement are opposing.

A magnetic line of force is a string with a positive *nuclear quantumized charge*,[1] which is an extension of a lineal link of protons and neutrons, rather than the protons themselves. "Axled" by a nuclear-quantumized-charge line, the flattened electrons in this string rotate. A nuclear charge is always

---

[1]   The term *nuclear quantumized charge* refers to the quantumized energy of the nucleus and is distinct from *nuclear charge*, which refers to the positive potential in the nucleus.

Fig. a

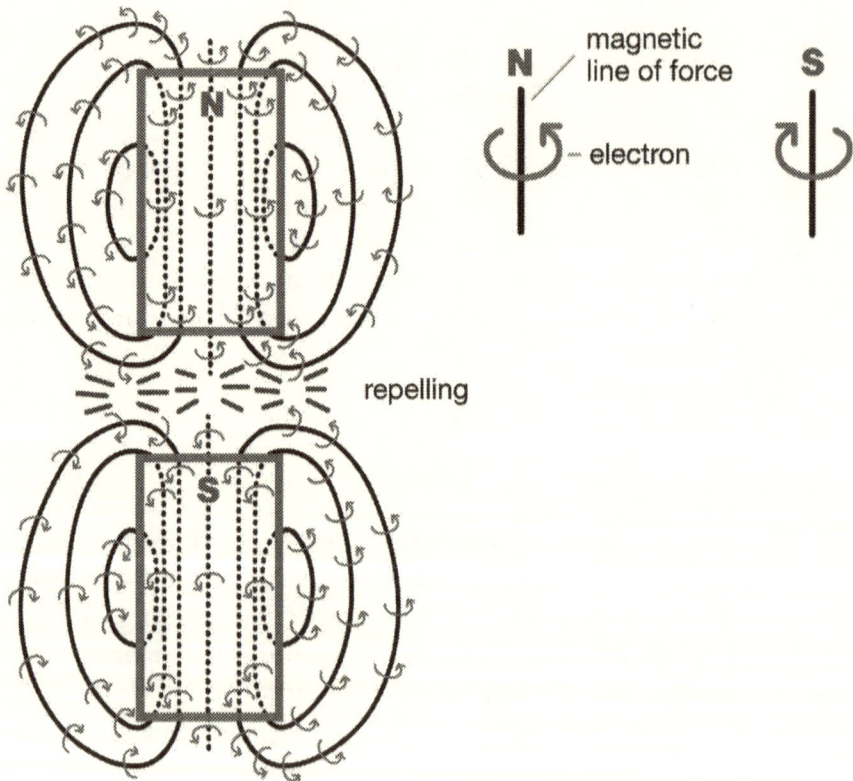

Fig. aa

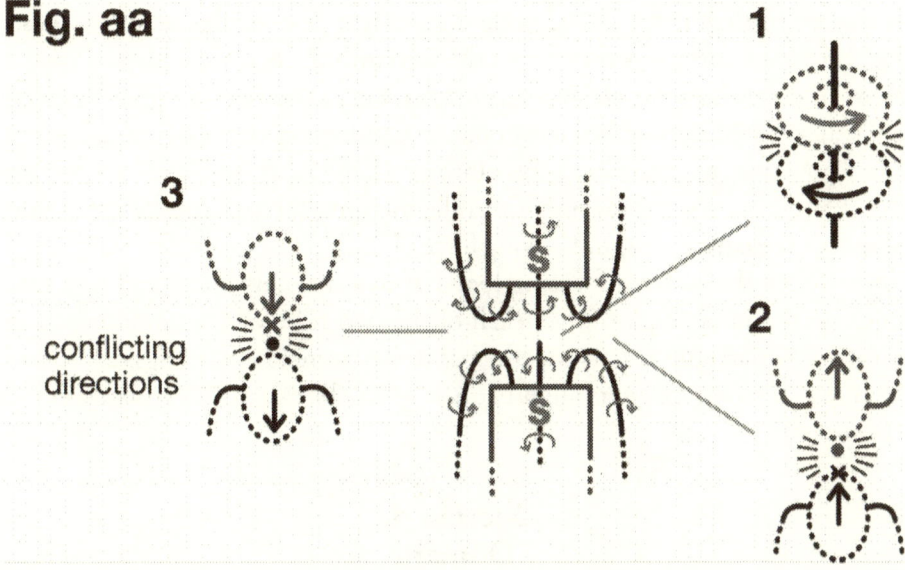

3

conflicting directions

1

2

perpendicular to the rotation of electrons. Similarly, a magnetic wave is perpendicular to an electric-current wave. If there is more than one lineal, nuclear-quantumized-charge line, there are an equal number of electrons to balance the positive electric charge; these rotate in layers around the combined, central nuclear quantumized charge. As a result, the magnet is extra strong.

Section 3.
Attraction

When the opposing polar ends of two magnets are placed in close proximity (Fig. b), they attract each other because their respective electrons merge when they move in the same direction at a particular point or phase. Fig. bb shows three aspects of how the electrons attract. Figs. bb 1, 2, and 3 show in detail how the electrons conform. Fig. c shows two opposing polar ends in a partial state of attraction before connection occurs. Figs. c 1 and 2 show how two magnetic loops fuse into one large loop. As Fig. c2 shows, the flip that occurs results not only from the merging force of the electrons, but also from the connecting and linking force of the nuclear quantumized charged lines. When the two conforming loops of two magnets break and fuse into one continuous string of force, each of the two sides of the attracting force pulls the two magnetic bars together, as if they were one bar (Figs. c3, 4, and 5). Thus, the two magnets align because of the lineal, nuclear quantumized charge.

Section 4.
Domain

The domains of a non-magnetized bar are jumbled because their molecules are jumbled, as shown in Figs. d and d1. The small circles represent the molecules, while the dashed lines inside the circles represent nuclearly-charged directions. In each molecule's atom, the electrons enwrap the nuclei, and their nuclear-charge direction may not correspond with those inside other molecules. Since the molecules are already in a crystalline state, this non-magnetized bar is characterized by nuclearly-charged directions that are switchable into a uniform direction.

The molecules in a magnetized bar are in a crystallinity in that their nuclearly-charged directions are arranged lineally. At the same time, their electrons revolve in a uniform direction, perpendicular to the lineal, nuclearly-charged lines.

Fig. b

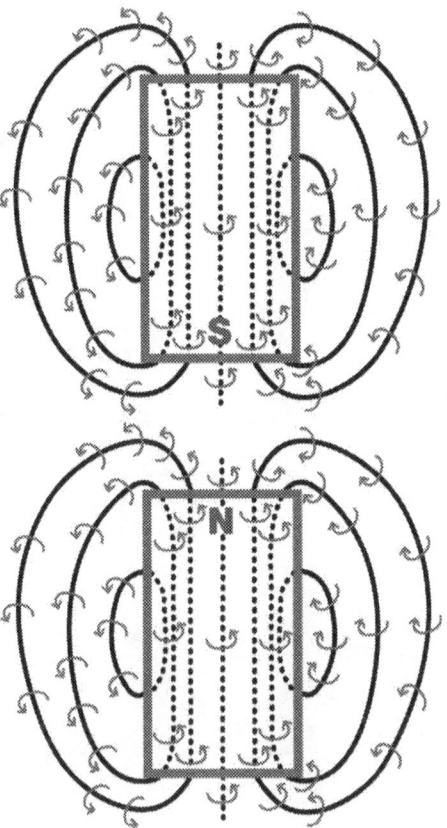

Fig. bb

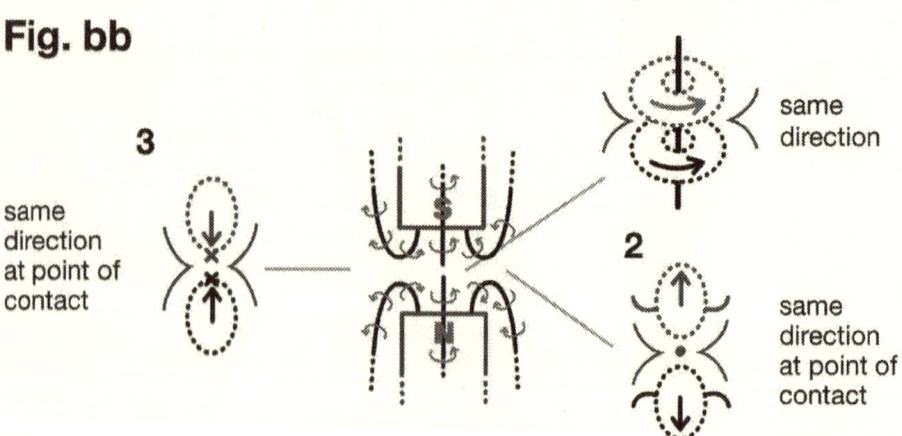

1

same
direction

2

same
direction
at point of
contact

3

same
direction
at point of
contact

Fig. c

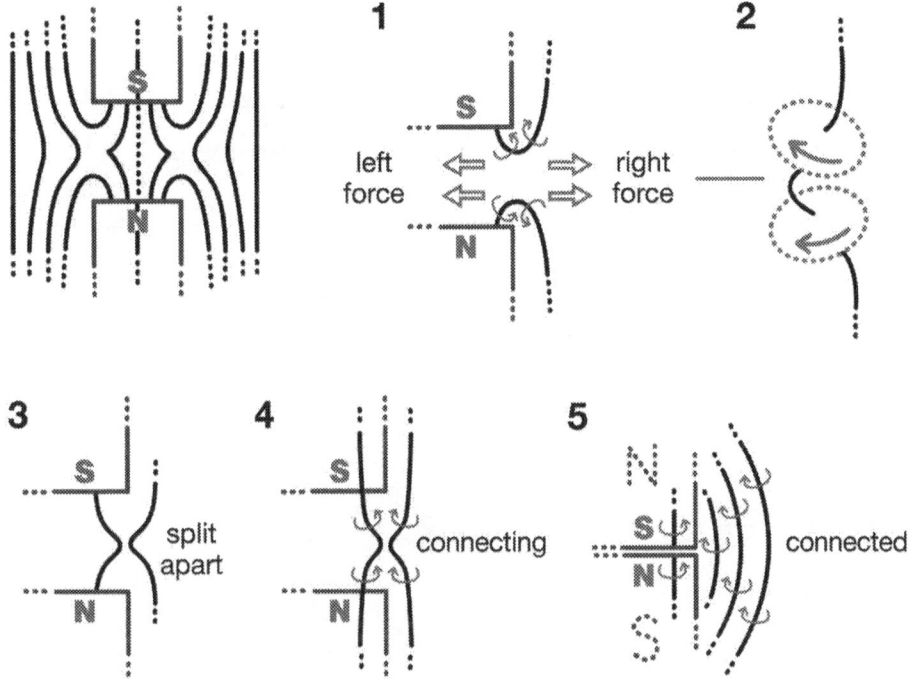

Fig. d

1

domains jumbled   crystallized pattern

=   oscillating electrons

2

domains aligned   electric charges aligned

=   revolving electrons

Section 5.
Induction

If a non-magnetized bar is stroked repeatedly by one pole of a permanent magnet in a single direction, then the permanent magnet acts as a comb that straightens the molecules in the non-magnetized bar so that they align their nuclearly-charged directions into one lineal direction. This act of connecting energizes the charge of the nucleus into quantumized energy, piercing through both ends of the enwrapped electrons, and forcing some of the electrons to revolve around the nucleus simultaneously. *Therefore, there is no magnetic flow direction of the field.* Therein, the nuclei connect their quantumized charges into a straight (or nearly straight) line, and all of the lines conform in parallel fashion in this induced state (Fig. d2).

In addition to this positive nuclearly-charged induction, there is a negative electron-charge induction, which occurs through rubbing electrons of a non-magnetic bar in one direction so that they revolve in a clockwise or counterclockwise direction. The revolving electrons will open both ends of their nucleus, expand their energy both outward and inward, and press out (by the electron's inward pressure) their nuclear quantumized charge.

Section 6.
Expansion

Fig. e shows that, when all of the electrons revolve around their aligned nuclei in a unified, rotatory direction, each electron in this string encounters its neighboring stringed electrons. As illustrated above (Section 2, Repulsion), in a magnetic bar, all of the electrons in a string unify their mutual force to repel neighboring strings. This is the expansion force inside a magnet, which squeezes out the nuclear quantumized charge, along with their revolving electrons, outside the bar from both ends as the extensions of the lines of field.

Section 7.
Contraction

As mentioned in Section 6, the magnetic lines of field extend from both ends of the bar simultaneously and continue to expand. Nonetheless, all of the lines tend to loop back elastically to their closest possible opposite ends. This forms a string loop and is one force to contract a magnetic field line, as shown in Fig. f 1.

Fig. e

2-D view   3-D view

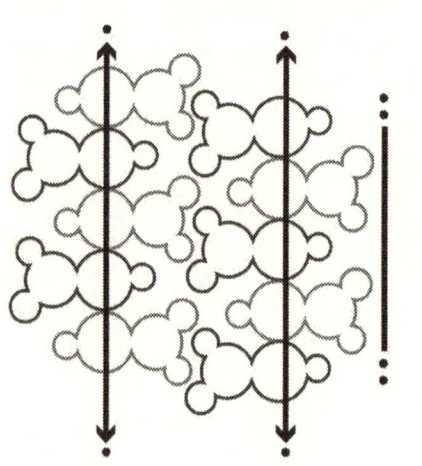

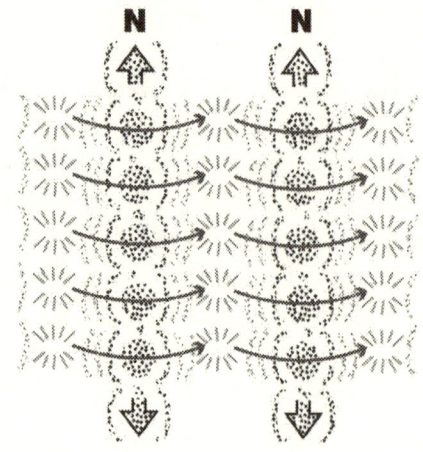

nuclear charges aligned   revolving electron kicks

Inside the magnetic bar, molecules align. Their nuclear quantumized charges pierce through the layers of electron barriers. The nuclear quantumized charges, which are connected, compress the energies of their oscillating electrons into revolutions. Although all of the revolutions around their nuclei may not be in a total state of equilibrium, their electrically charged equilibriums are sustained.

Still, the excessive energies resulting from the nuclear quantumized charge and the electrons' own revolving and expanding energy require more space. Therefore, they continue to expand, simultaneously extending outside the bar at both ends. Once outside the bar, they continue to expand. Thus, the magnetic force has no flow direction.

Fig. f

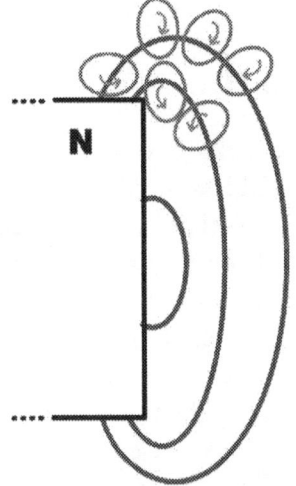

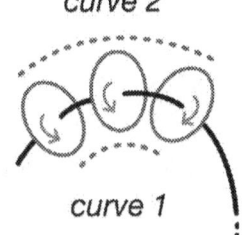

curve 2

curve 1

Still another contraction force accounts for the magnetic lines of force outside the bar forming a string loop. This is the power of the rotating electrons in the same string outside the bar. The space outside enables these lines to move further apart, thus helping to bend the lines into loops. Since each line is itself directional at both ends, the electrons in the string aid in the bending. Fig. f 2 shows that the rotating electrons are closer to each other on the concave side of the string (curve 1) than on the convex side (curve 2); accordingly, their merging forces are stronger along the inner side of the curve. Thus, most of the magnetic field lines are forced to curve and loop back to their opposite ends, forming completed loops once inside the bar.

This contraction acts as the force of attraction. Along with the nuclear quantumized charged lines, the rotating electrons in each string create the pathway that enables the bending lines from both ends to align and meet.

Section 8.
Impact and Heat

Striking or heating a magnet impedes the revolutions of the electrons inside the magnet. In some cases, it could even cause the lineal nuclear quantumized charge to break into sections—that is, demagnetization.

During magnetization, high heat will impede the electrons' revolving capacity, encouraging them instead to vibrate with excessive energy (oscillate). However, this factor, in turn, will weaken the repulsion between the electrons of neighboring lines. Conversely, it will excite the magnetic molecules to align closer to other lines in the array, causing them to enforce the magnetic force when the magnet cools. Upon cooling, some electrons dovetail into the empty spaces or holes of the atoms, while others revolve around their nuclei. This is how a magnet forms.

Section 9.
Material Features

In various materials, the molecules are arranged in different patterns; thus, their magnetic reactions may differ. The main factors are related to their nuclear angular momentum and/or the direction of their electro-revolution.

Fig. Ferro shows that, when a magnetic bar is close to a ferromagnetic bar, it is the nuclear quantumized-charge of the magnetic bar piercing through the ferromagnetic bar that causes induction and attraction.

Fig. Ferro

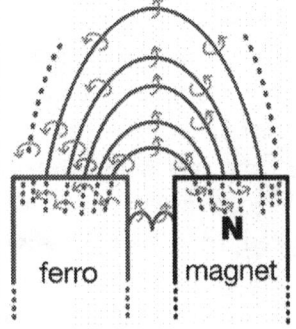

Figs. Para 1 and 2 show two examples that might distort the magnetic force lines. Fig. Para 1 shows that, in a paramagnetic bar, the nuclear-quantumized-charge arrays are linked in spirals. Fig. Para 2 shows the nuclear quantumized charge lines curved into non-uniform directions.

Fig. Self-Repel shows that a self-repellent magnetic bar has a distant mode of nuclear quantumized charged lines. When induced by a magnetic bar, only those revolving electrons in one nuclear quantumized charged line are affected. In this way, the electrons throughout the string simultaneously rub the neighboring lines' electrons in an opposing, revolving motion because of the too large, fixed intervals between the molecules. Even though, string by string, their electrons revolve in opposite directions, these electrons still cannot merge at the points or phases where they touch. However, once outside the bar, the lines couple into one rotating direction, while the neighboring couple is forced to rotate in an opposite direction. Thus, those contrary coupled lines repel each other.

Fig. Non shows several nonmagnetic metal bars placed alongside magnetic bars. Some nonmagnetic materials allow a magnet's nuclear-quantumized-charge lines to penetrate sectionally but discontinuously. Other nonmagnetic metal bars allow electrons to revolve but prevent the nuclear quantumized charge from piercing through. Still other nonmagnetic metal bars are structurally fixed; neither their nuclear quantumized charge nor their electrons can be induced. In still other cases of nonmagnetic metal bars, even if everything else is inducible, their neutrons always slide to meet the inducing lines, thus blocking them from further action.

Fig. Dia shows that the nuclear-quantumized-charge angularities in some diamagnetic materials are ever changing; yet, their electrons meet no resistance from being induced to revolve. On the other hand, some nuclear quantumized charges are fixed in a non-lineal array; however, their electrons, through the magnetic bar's revolving electrons, are induced to revolve freely, expanding to form smaller, individual magnetic loops.

Section 10.
Super-conductor

As Fig. Super 1 shows, when a magnetic bar is placed upon a superconductive material, extremely strong reactions will occur. As the magnetic bar approaches the superconductor, even the outermost field lines of the bar, along with their rotating electrons, will agitate ("rub-induce") the surface electrons of the superconductor.

Fig. Para

1

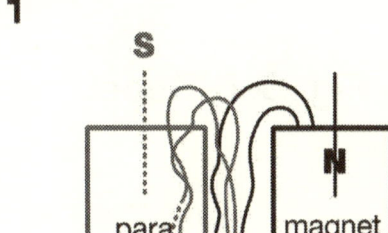

2

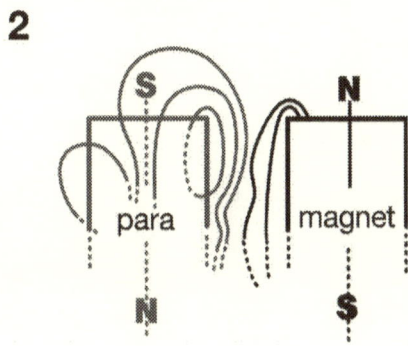

Fig. Self-Repel

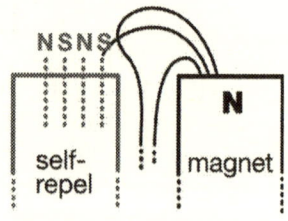

Fig. Non

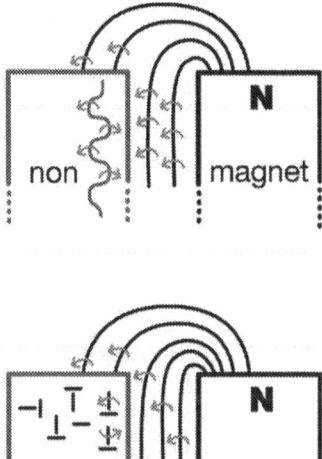

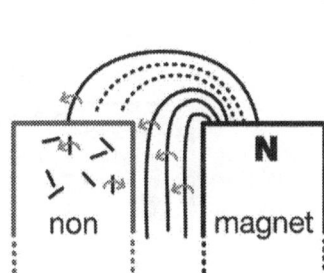

Fig. Dia

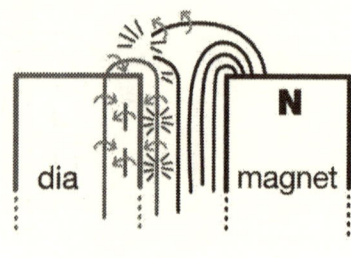

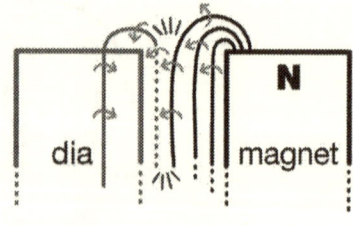

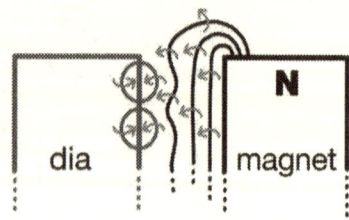

Fig. Super

1

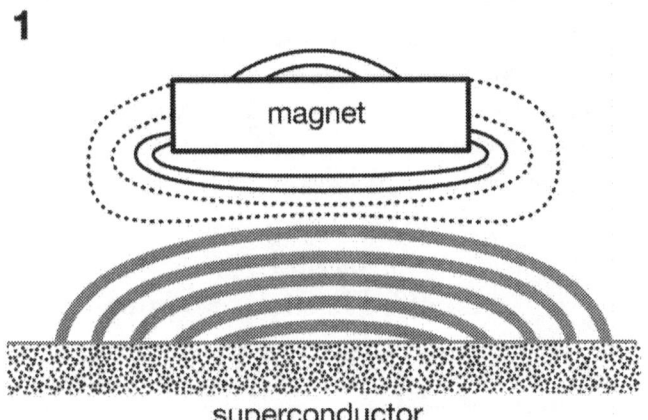

superconductor

2

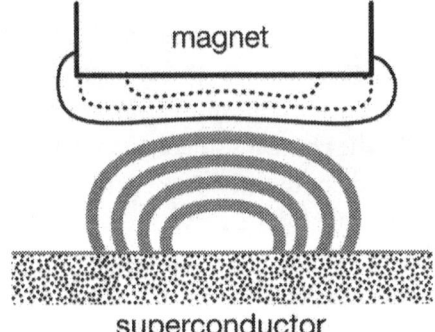

superconductor

In a superconductor, most of the directional, nuclear quantumized charges are extremely turnable, meaning that those charges no longer retain their lineal connective character. Instead, those nuclear quantumized charges are "super-connective;" they immediately bend to form loops, either small or large, to oppose the inducing field strings of the bar. At the same time, the super-conductive electrons revolve super fast, which greatly enhances their flattening, compaction, and expanding.

Flattening enables their axial nuclear quantumized charge to bend easily (i.e., to form the smallest possible loop). Expanding enlarges the space between neighboring lines of the magnetic field. Combined, these two factors form a super-strong magnetic force to repel the inducing magnetic bar.

As Fig. Super 2 shows, it takes only one field string to stimulate more than one of the superconductor's looped strings. Thus, the repelling force of the superconductor is always stronger than the magnetic force of the bar.

Section 11.
Earth's Fictitious Magnet

The Earth's ellipticity seems to indicate that its fluid originally had an equatorial-plane directional flow out from the center axis of the Earth's spin, separating into two opposing circulations, returning via both poles, to the central axis. This nearly symmetrical circulation began to form the Earth's fictitious magnetic bar. And, at that time, the Earth's spin axis was the same as the fictitious magnetic bar. Over time, iron and nickel particles,[2] through the force of magnetic attraction, gradually consolidated into a solid, center core, strengthening the fictitious bar's magnetism. Friction of the center core with the surrounding fluid layer caused the electrons of the center core to revolve clockwise since the center core spun from west to east.

Yet, in the rod generally, the electrons still oscillated because of excessive heat energy. Before the magnet had completely formed, the polarities of this rod had already been determined, mainly by the gravitational field of this solar system, which has determined the west-to-east spin of the Earth. Fig. g, a cross-sectional view of a magnetic bar, shows how the angular momentum of the electron charge is balanced. In this case, all of the electrons revolve counterclockwise inside the bar; thus, its total angular momentum inside is counterclockwise. However, in the field, all of the electrons rotate clockwise. Conversely, as Fig. Earth1 shows, since the Earth turns from west to east, the Earth's field has a west-to-east

---

[2]   Since the gravitational force decreases toward the Earth's center core, other heavier
       substances did not accumulate in the core.

Fig. g

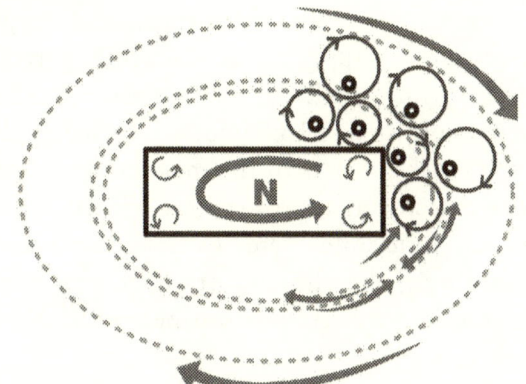

cross-sectional view
of magnetic bar

The outer electron angular momentum is always
greater than that of the inner counterclockwise
direction. The same pertains inside the bar. Thus,
the total angular momentum outside moving in
one direction balances the total momentum inside
the bar moving in the opposite direction.

direction of angular momentum. This momentum results in a counterclockwise total angular momentum of the Earth's magnetic field. All of the electrons in the field rotating in this direction are affected, while inside the Earth's magnet, the free clockwise revolutions of the electrons are mostly impeded by the high heat. Thus, the electrons oscillate as they slowly revolve. The electrons inside a magnetic bar and those in the same plane outside the bar must revolve in opposite directions; otherwise, no magnetic force field can form.

Fig. Earth 1 shows the polarities of the Earth's magnet, and Earth's completely formed electro-magnetic circulation system. Little by little, the Earth has been losing its electrons as objects have cut the strings of the magnetic field. The chemical reactions of hydrogen and oxygen released a large amount of electrons to supply the axis' magnet, partially compensating for this loss. At that time, the Earth rotated faster (it is said that 1.2 billion years ago, one day consisted of 18.8 hours). The geographic North Pole was centered at the south polarity of the magnetic pole, and the geographic South Pole was centered at approximately 130 °E and 65 °S. Originally, the Arctic and Antarctic zones were positioned differently than they currently are. Accordingly, the climates of geographic zones were different. For example, southern Siberia was warmer, while North America was much cooler.

Along with continental drifting, continual changing and reforming of the Earth's crust, together with the force of water (from ocean currents, tides, and the Earth's rotation) combined with the change of the Earth's angular momentum to cause the Earth's land mass to become unbalanced. The imaginary axis of the Earth's rotation began to swing away from its original axis, to which its magnet was anchored, until arriving at its current position. In the future, the swing could cause the axis to either return to its original position or spiral away. Along with these above-mentioned factors, the moon's pull on both the Earth's crust and mantle and the rotation of the Earth's fluid under the mantle slow the Earth's rotation. This implies that the spin of the Earth's core also slows down.

The slowing of the Earth's rotation, in turn, slows the inside circulating fluid and the re-supplying of electrons to the Earth's core. Figs. Earth 2 and 3 show another factor related to the weakening of the Earth's magnetic field force. Fig. Earth 3 shows consolidated masses of ferromagnetic materials,

Fig. Earth

1

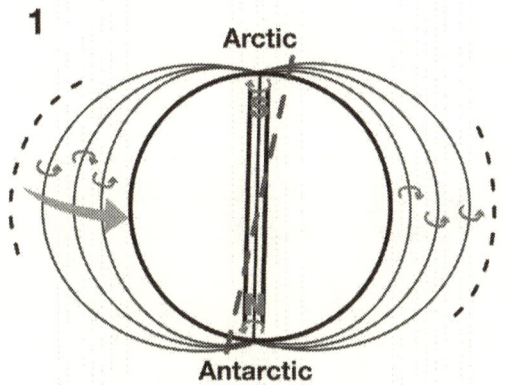

Arctic

Antarctic

2

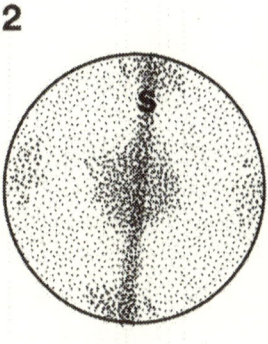

newly formed magnet

Because of the effects of heat, electron revolutions inside the bar or rod are just a mean momentum. Since the substance between the magnet and the mantle is viscous, the direction of its total electron momentum should be less significant because those directions are ever changing. Moreover, they are oscillating, rather than revolving.

Generally, the electrons outside the Earth's crust will rotate in a counterclockwise direction. However, the fluid swirling back through both poles turns symmetrically and aids in forming the Earth's magnetic polarities. As shown in Figs. a and b, clockwise turning becomes magnetic "S," and counterclockwise turning becomes magnetic "N." Thus, the Earth's magnetic polarities have been fixed.

3

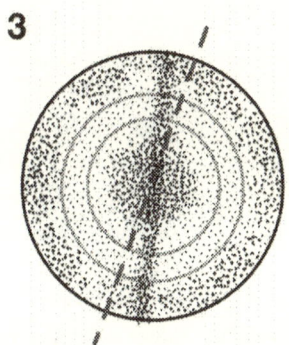

Some ferromagnetic materials tend to be spun to the surface, while most are attracted to the center magnet.

which create their own smaller magnetic loops inside the Earth, thereby weakening the outside field.[3]

Section 12.
Conclusions

a. The directional turn of the electrons will determine the magnetic polarity. While the nuclear quantumized charge does not determine the polarity, its rotating electrons can still affect the directional bend.

b. The directional turning of an electron depends not on the electron itself, but on perspective. That is, clockwise turning viewed from one side is counterclockwise turning when viewed from the opposite side.

c. The magnetic line of force has no flow direction; rather, the line of force extends from both ends of the bar into the air simultaneously.

d. The magnetic line of force is neutral and can penetrate most materials. (Two positively charged balls repel each other because their field lines have energy potential [not rotating electrons]; however, magnetic field lines do have rotating electrons.)

e. The magnetic lines of force are derived from the nuclear quantumized charges, which are perpendicular to the fields of their electrons' rotation.

---

[3] When the mantle is uneven on the surface, it will also be uneven under the surface. This will impede and distort the flow of fluid and cause an ionic result. The ionic result under the surface will induce ionization in the air. That is, a disturbance of fluid movement occurring underground has an effect above ground. For example, if the air above ground is full of water droplets, then the gathering of them through ionization may gradually form a hurricane or a similar natural event (aided by temperature, winds, etc.). Perhaps the ionic status could be controlled by applying liquid electrons (i.e., condensed electricity) to form clouds or to collapse a forming hurricane. Another case worth mentioning is as follows. When a heavy layer of clouds forms in the air, even a small de-ionization may trigger a chain reaction that collapses the entire layer of clouds. Suppose, for example, that an airplane is flying under this heavy cloud layer. Since the airplane will, more or less, carry a charge and will cut the magnetic field strings in the air, some released electrons will be attracted to the clouds. A sudden "feed" of electrons can trigger an abrupt collapse of the clouds. Then, a funneling of both the surrounding water droplets and the air may directly impact this airplane. Therefore, small airplanes should avoid flying directly under heavy layers of clouds.

f. Moreover, the magnetic lines of force consist of the field surrounding the bar and the connected looped strings of nuclearly (quantumized) charged extensions with their rotating electrons inside the bar.

g. Lightning rods are a minor factor in reducing the Earth's reserve of electrons and weakening the Earth's magnetic field of force. Therefore, it is suggested that they be equipped with a sensor switch to avoid unnecessary loss of electrons.

h. Other factors contributing to the weakening of the Earth's magnetic field of force are consumption of the Earth's own energies (resulting in loss of electrons), thickening of the mantle (mantle thickens as electrons are lost, which, in turn, impedes the electron "feed" to the Earth's core), less underground water (without water, conduction is impeded), the moon's pull (which releases some of the Earth's electrons), and a host of manmade infrastructure and equipment that interferes with the circulation of the electron "feed" to the Earth's magnet (e.g., railways and tracks, buildings, ground and air traffic, and power lines).

i. Although the Earth's magnet appears fixed, recent observations suggest periodic reversals in its polarities. (This phenomenon will be discussed in Part IV.)

j. Had the temperature of the Earth's center core not been so high (estimated at approximately 6,900 °K), the electrons of the Earth's center core would have revolved more freely, and the magnetic field of force of the Earth's surface would have been so strong that the free movement of any ferromagnetic material would have been impeded (e.g., a steel blade could not have been swung freely).

PART II

Photons: Wave-Particle Parallelism

Section 1.
Wave and Particle Features

WHEN A BEAM of light is directed onto a smooth surface, its incidence angle equals its reflection angle. However, when a beam of light penetrates a substance whose density is greater than the medium of the air, the light will refract. When light, which travels in a straight line, passes through a slit about the same size as its wave length, it diffracts. When light is beamed through two neighboring slits, the diffracted waves interfere with each other. When a beam of white light passes through a prism, it will bend and split into the visible color spectrum. When a second, inverted prism is joined to the first, the color spectrum will merge back into a thin beam of white light upon leaving the second prism, which again travels in a straight line. The density of the prism, which is higher than that of the air, causes the speed of light to slow, separating the light into different colors. Since this is the case, why does adding the second inverted prism cause the spectrum to combine into white light? Why doesn't the light slow down even more?

Section 2.
Portrait of a Photon

A photon has its center energy quantity and surrounding energy field, which combine uniformly, carrying positive and negative electrical charges. The density of the center is higher than that of the surrounding energy field. Photons have different energy quantities, frequencies, and energy field sizes, depending on where they are generated. In general, the closer to the nucleus a photon is produced, the higher its energy quantity, frequency, and energy field (such as x-rays or ultra-violet rays). Conversely, a photon generated farther from the nucleus has less energy quantity, a lower frequency, and a smaller energy field (such as red light or infrared light). As a photon is being formed, part of it is already leaving the source; thus, the shape of the photon's center energy quantity is not round, but elongated. Because of the spin of the generating source, the photon travels a spiral path. Thus, a photon carries its own wave field.

Section 3.
How a Photon Travels

When a photon is pulled from its source, its initial movement is sluggish; it takes on a spring coil shape, and its motion becomes helical. The faster the spin speed of the source, the more energy quantity is pulled out and the higher its frequency. Conversely, the slower the spin speed, the less energy quantity is pulled out and the lower its frequency; as a result, the photon's particle and wave fields are smaller and thinner and the shape is more elongated. Thus, the photon's energy quantity size and frequency both depend on the strength of the spin speed. Each photon has its own power of coherency to maintain its respective core and wave field. As photons are generated from their source, they radiate out in a straight direction; their respective wave fields start to expand, and mutually repel each other. The repulsion effect of mixed frequency, such as white light, causes greater interference and thus more dispersion. Photons of the same frequency, expanded to a certain degree, will stay together, as in a laser beam. As the photons expand, they maintain a straight line. Because the photon itself has a positive and negative charge, it will be affected by the curvature of the gravity field, and it will bend. (This phenomenon will be explained in Part IV.) If the source is moving in the opposite direction of the photon generated, then the length of that photon will be stretched out further and its frequency will be lower. That's a part of the Doppler effect. If, on the other hand, the source is traveling in the same direction as the generated photon, then the photon will be compressed and its frequency higher. Although the photon's frequency can change, its speed will remain constant since this is derived from the electrical charge's repulsive force.

Section 4.
When Interference Occurs

Many experiments have been conducted to prove the photon's duality feature. For example, when a board with either two thin slits or small holes is placed between a light source and a screen, the photon reveals its wave feature. The reason is that the board, which has its positive and negative field, interferes with the photon. The positive and negative charges along the edges of the slits or holes partially absorb the photon, dissembling the particle into an energy wave. The nature of the interference corresponds to the matter through which the photon travels. If two small slits, the screen will show stripe-like interference; if two holes, circular interference. Thus, a photon's nature is solely that of a particle; once interference occurs, as evidenced by the wave, the photon no longer exists.

When a beam of light is projected through a polaroid lens, the light becomes polarized. That is, the well-ordered positive and negative crystal construction inside the lens will prevent much of the photon's energy from being absorbed; the spiraling path of the light will be narrowed into an elongated one, close to that of a sine wave. The angle of elongation will twist slightly if bio-rotation or bio-refringence occurs. This explains why polarized light can be used to detect the existence of life forms.

Section 5.
General Phenomena:
Reflection, Diffraction, Interference, and Refraction

When a **spiraling** photon encounters a smooth surface, the incidence angle equals the reflection angle. Even though the photon spirals, it still travels in a straight line and has the particle nature. The phenomenon of diffraction occurs when light passes through a small slit, the size of which about equals that of the photon's wavelength; it no longer travels in a linear direction, spreading into curving waves. A beam of photons traveling past a slit is comparable to an audience leaving a concert: The audience members congregate as they move toward the exit; once outside, they go their separate ways. Similarly, when photons encounter a larger slit, they pull in other photons. As their own surrounding fields are impeded, they cohere to adapt to the surrounding pressure. Once outside the slit, they again expand.

Polarization of light is elliptical.

Interference happens when two or more adjacent diffractions occur.

The cohesive force between photons causes them to become a collective field. At the same time, a repulsive force causes them to push each other away because a photon carries both positive and negative charges. Before the photons radiate out, they cohere; once traveling, they expand. As photons pass round the edge of some material, such as a thread, some are damaged and absorbed. The undamaged photons expand to fill the now empty space, which causes them to turn. This expansion explains, in part, the diffraction phenomenon. Most materials have a surrounding positive and negative field; because of its miniscule size, a photon passing by this field will be strongly affected.

Refraction is discussed in the section that follows.

Section 6.
Refraction and Prisms

When a beam of white light travels through the air, it contains photons with different frequencies. Once the beam enters a medium with a different density, it will, according to its entering angle, bend toward the side with more molecules. For example, when a beam of light penetrates a double-sided, transparent medium, such as a glass pane, if it is not in a normal angle to the surface, the light will bend toward the side with more molecules. Thus, refraction occurs.

Similarly, when a beam of white light penetrates a prism, the beam tends to bend more toward the side with more molecules. The higher-frequency photons, such as violet light, will touch the molecules many more times and thus bend, or actually curve, increasingly toward them; while the lower-frequency photons, such as red light, will touch the molecules fewer times; naturally, they will bend or curve less toward them. When the spectrum of light emerges from the prism, it will bend or curve a second time not because of a density difference since the air's density is less than that of the prism. Rather, the reason is that the photons again tend toward the side with more molecules. The degree of curvature depends on the photon's frequency. Again, higher ones bend more; lower ones less. Therefore, the spectrum appears wider. When an inverted prism is placed up against the prism, the colors bend or curve back, re-combining into a beam of white light. Thus, refraction occurs.

The bending of the beam of light appears angled in a straight direction; however, it is actually a continuous curve. And once all the photons stop curving, they continue traveling in a straight line. Because the area of occurrence in the curve is too small to be noticed, the gradual curvature into the medium has been mistaken for an abrupt, bent angle.

Section 7.
Summary and Conclusions

In conclusion, a photon's quantity and frequency depend on where the photon is generated in relation to the nucleus. Photons produced closer to the nucleus have a higher quantity and frequency, while those generated farther away have less quantity and lower frequency. The two ends of the elongated photon have opposite but balanced electric charges; thus, the photon is considered to be neutral. A photon carries its own wave field. This field has its cohesive and repulsive forces because of its different electric charges. If part of the photon's field has been attracted by a stronger charge, then the photon will follow and be pulled into the field of the stronger charge; that is, either absorption or destruction will occur.

The photon has its mass or quantity even when traveling (it has been believed that a photon is massless). Naturally, it will be affected by the gravity field. Before a photon has been ejected, it has its intrinsic energy. During traveling, its initial momentum derives from that energy. When the photon has been stopped, the momentum is compressed back into intrinsic energy. Once the photon is released, this intrinsic energy will again spring out, resuming the speed of light. According to the photo-electric effect, light can generate electrons. Do the photons turn into electrons? If so, where does the remaining positive charge go? Perhaps the remaining positive charge will be consumed into energy, along with the neutrinos; if this is the case, the photo-electric effect will not only deal with the electrons but with the neutrinos as well. An extreme photo-electric effect possibly will generate positron. The photon derives its initial speed, cohesive and repulsive forces, and spiraling momentum from both the positive and negative charges. The spiraling momentum enables the photon to continue in a straight direction once the interference factor is over.

The λ, wavelength of the light, depends on how close the photon is to the center of the atom. The shorter the radius, the stronger the nuclear angular momentum, the higher the volume of the photon, and the higher the photon's amplitude and frequency. Photons have the cohesive and repulsive forces because they are composed of positive and negative charges at the same time. Before they travel, photons cohere; when they travel, they expand. If the photon were to lose its volume during traveling, its frequency would decrease. If the photon were to diffract or lose some of its outer wave field, the light would age, both frequency and quantity would decrease, kinetic cohesiveness would decrease, and the photon would self-quench into neutrino-like elementary particles.

According to the Theory of Relativity, the speed of light is the fastest speed in the universe. If a photon were to be transmitted from the side of spacecraft whose traveling speed approached the speed of light, that photon would not combine its own traveling speed with that of the spacecraft in an oblique angle. Rather, it would maintain the angle directly from its source. The photon would not have been affected by the spacecraft's speed; it would have been affected only by its source repulsive charge and direction.

A laser beam contains photons of nearly the same frequency. Thus, its photons do not expand their carried fields laterally; rather, they possess an affinity for each others' fields and connect into *non-continuous* streams as a flow of light.

Because the photon carries its own wave field, its impact is buffered and, via its oscillation, the receiver's range of recognition is enlarged; otherwise, photons would collide with and damage each other. In addition, they would damage the receiver, such as the human eye. Thus, the photon's wave field allows it to travel without cessation.

PART III

On Color Force and Color Charges

Section 1.
Asymptotic Freedom and the Color Force

BECAUSE ITS ELECTRIC charge is incomplete, a single quark can neither attract nor repel other particles. Likewise, it cannot form a minimized field of force around itself. The quark's easily broken field needs an incomplete, opposite charge to match. Once two quarks of opposite charges meet, their color force is immediately enforced. This will then attract a positively or negatively charged third quark; together, the three quarks form into one group, either a proton or a neutron; simultaneously, the quarks' fields balance their different color charges in presumably three concentric membrane-like layers within which the quarks are self-confined. The thick, less voluminous innermost layer is surrounded by a thinner, middle layer with clear color charges; this layer, in turn, is surrounded by an even thinner, totally complete outer layer, with the strongest color charges, which reinforce or enhance the color force to the utmost. When the quarks are close together, the color charge is weak and the quarks behave similar to free particles. This phenomenon is known as asymptotic freedom.

As the quarks move apart, the repulsive force derived from their color charges becomes stronger. Quarks that attempt to penetrate the barrier of their own mutually-structured membranes cannot escape. As the inner membrane is forced closer to the second and third membranes, the repulsive forces between the membranes strengthen. The membranes, which originated from the quarks, repel the quarks closer together, and they resume moving freely.

Section 2.
Construction Beyond Quarks

The completed baryon—either a proton or neutron—releases its field into a layer of an extra-nuclear field, whose structure is analogous to a platelet-like membrane. For the neutron, the outer layer consists of its most negative charge. For both the proton and the neutron, a dense neutral field forms beyond the baryon's outer layer, which gradually extends to a less dense, neutral field. The more baryons that combine into a nucleus, the more structured layers there will be. These so-called membranes have phasial contrasts, whose extra-nuclear interactions are phase waves. Because they are of smaller dimension, the innermost layers have the highest density and strongest force. Successive layers farther from the center have progressively larger dimensions, less density, and weaker force. The stronger of any two successive layers exerts a repulsive force toward the weaker one. Layer by layer, the gaps increase until all layers balance the baryons' color-force fields; that is the boundary.

Once the nucleus of the baryons is completed, the color-force field oscillates. This oscillation is based on the interference force of other oscillations, which might turn into spin. In other words, oscillation and spin are interchangeable as both are a form of potential energy. After the baryons are formed, some positive charges will penetrate out to and/or through the median fields, which attract electrons.

The color-force field of the nucleus does not impact the electrons directly. Rather, the positive charge information passes from the nucleus to the electrons through a medium, a neutral field consisting of neutral energy with less mass. Accordingly, those neutral energy fields are part of the source from which mass is derived. (Currently, they are not recognized as existing as mass.) The closer the layer is to the nucleus, such as the K layer, the higher the density, the smaller the field but stronger the positive force to attract the electrons. The larger layers farther from the nucleus could have less density and weaker energy, but the greater dimension can hold more electrons. The neutral field between density layers acts a barrier to prevent the electrons from getting too close to the nucleus.

Section 3.
Nuclear-binding Energy: Strong (Color) Force

Between two protons, there is no binding energy or exchanging force; there is only repulsive force. However, between a proton and a neutron, there is strong binding energy, such as in the hydrogen isotope deuterium. When a second neutron is added to the proton, the proton reacts to the neutron's negative outer field by attracting an extra neutron, such as in tritium. When two protons attract two respective neutrons and the resulting pairs encounter each other, they become double-binding, which causes the protons and neutrons to stop spinning; however, they oscillate (e.g., the α particle). This is the strongest form of nuclear binding. In Section 1, it was mentioned that three binding quarks have their excessive outside field. Similarly, there are neutral neutron fields outside the nucleon.

When pairs of one proton and one neutron increase, their matching proportion will diminish. Therefore, they need more neutrons to embed into their gaps because the neutron has the delivery function for the positive or negative charges. Therefore, the strong force is the attraction force that passes between a proton and a neutron through their neutral fields; they never touch, but squeeze out their fields and share a mutual outer field, which has a combined charge. With regard to the mesons and gluons pushed out from the accelerators, these are the fields between protons and neutrons; that is, the neutral quantities of the fields. When these quantities of the fields have been pushed out, the binding bridges between the protons and the neutrons or between two neutrons are neutralized fields, which are combined fields from the protons and neutrons. They do not exist as oscillating particles; rather, they are the quantities of the fields—that is, they serve as the medium or bridge—through which the color forces between protons and neutrons can bind strongly. Once these quantities of the neutral fields have been pushed out, the protons and neutrons or the neutron pairs will split. The neutral fields that have been pushed out are recognized as mesons or γ rays.

When protons increase in number, the position, angle, and some complexity of this matter increase. More neutrons are required to maintain the position and angular momentum in equilibrium. The magic numbers of either protons or neutrons—2, 8, 20, 28, 50, 82, and 126—are well known. Elements with these numbers will be very stable. Thus, the solid geometrical position is more symmetrical. Otherwise, the symmetrical position in the atom's nucleus will be unstable. In that case, the nucleus would gradually become either an isotope to be balanced by pushing out more neutral fields, α particles or γ rays, or simply require more neutrons to balance or embed in those gaps.

Having too many neutrons in the nucleus does not mean that the color forces are balanced; rather, it means they consume excessive positive charges, which causes the negative charges to split. This reaction is known as nuclear fission. Fusion, on the other hand, results when a stronger binding congregation squeezes out even more quantities of their mutual neutral fields.

Section 4.
Weak Force (Color Charge)

The weak force is caused by the nucleus being in a metastable state of equilibrium. Because the nuclei still retain their excess neutral energy fields with the nuclei's intrinsic angular momentum, those weak forces will be circularly released. The unbalanced nuclear resonance continues the radioactive activities until nuclear stability is reached. This unbalanced state still involves positive and negative charges. The proton and neutron, in an ongoing state of intercharged oscillation, have their respective fields' capacity to readjust their fields to achieve balance. The excess fields, including α particles, will be sacrificed and released from the nucleus. Those forces derived from the electrical charge or repulsive force will therefore release β particles. Its central field—that is, the γ wave inside the nucleus—acts as a negative force, which has its quantity. Once released, along with α particles, its presence is neutral and it has its quantity, just as a photon does. However, neither the γ ray nor the photon is massless. Both exist in an energy form; therefore, all wave and energy form is in a "mass" state. Similarly, all matter is in an "energy" state. Then, both matter and energy are in an overall energy state; that is, they do not need to change to become energy; only the form of matter is in a very stable energy state. Thus, all existence is in an energy state.

When a chemical reaction occurs, the oscillating effect that results from the change of electrons will still vibrate through the bridge of the neutral fields into the nucleus or nuclei. The degree of the responding oscillation depends on the intensity of the vibration or revolving of the electrons and the related neutral fields. Only, in some cases, the oscillations are too miniscule to be observed.

Section 5.
Forces Derived from the Three Charges

Two quarks with opposite color charges are incomplete; though they attempt to compensate each other, the saturated charge for each quark is of extremely short duration, causing the quarks to exchange the charge. This intermittent intercharging of the conjugated charged pair causes oscillation. This oscillation, in turn, requires charge compensation; thus, the pair attracts a third quark, which completes the charge, and causes the baryon to start to spin. Thus, when two quarks are joined by a third one, they complete their full color charge, and oscillation immediately becomes self spin.

The excess positive charge has energy to expand to become elongated and shrink to become thicker; while the negative charge has a string nature, whereby it can laterally expand and contract and expand its circular dimension into revolving energy. Because the negative charge of the neutron is at the outer circle, when it encounters a proton, it will surround it through the neutral medium. That is the so-called strong or color force. When a proton encounters a second neutron, the second neutron will be attracted to the proton. At that moment, however, the binding power of the two neutrons and one proton is reduced by half. The two neutrons can again attract another proton to bind strongly and continue on. That is, neutrons outside the negative charge always have excess weaker binding to the proton. As can be observed in some nuclei, the number of neutrons could be more than the number of protons.

When conjugated energy gradually reaches a fully balanced degree, it has achieved its highest boundary. Any excess energy that overflows this boundary or layer will precipitate an abrupt drop to a much weaker, but uninterrupted, energy level. At that moment, the higher-level layer becomes a major influential energy. Ordinarily, influence from the lower-level layer will have little effect on the higher one. However, an enormous amount of lower-level energy can still cause the higher-level energy to respond. Conversely, a higher-level layer of energy can easily affect a lower-level one. Therefore, when the core of the nuclei oscillates, its outer electron layer will be compressed from the center outward, and the electrons start to revolve. When the core of the nuclei has a low oscillation, the outer layer of electrons attaches to this outer field to oscillate.

A positive charge has its affinity with another positive charge and a negative charge has its affinity with another negative charge if the direction of the oscillation or spin is the same; otherwise, they repel each other. If their frequencies differ, they also repel each other. If their frequencies are similar, they attract each other. The nature of the positive charge is to expand linearly. If affected by another force, it can be bent into a closed ring, unless otherwise broken. The nature of the negative charge is to expand into oscillation or revolving.

Because the positive and negative charges differ in their features of oscillation, different elements can easily compensate each other's oscillation into the same frequency. For example, a crystal has its positive and negative polarities, which are balanced to form the same frequency. The linear feature of the positive charge and the lateral revolving feature of the negative charge transform atoms into matter.

A neutron's complex functions result from its positive, negative, and "neutral" charge features. The neutron field enwraps that of a proton and stretches or contracts with it. The neutron field also insulates the proton from the electron. In addition, the neutron field prevents the electron field from coming too close and thus parallel to the proton field. In this way, the perpendicularity of the positive and negative charge fields is maintained by the field of the "neutral charge."

The γ ray, meson, and gluon are examples of part of the weaker force. They are generated from excess neutral charge field, which the positive charge field can no longer hold and must release.

Section 6.
Conclusions Regarding the Three Charges

When charge differential fields tend to become charge symmetric fields, this type of quanta transition causes a range of phenomena. These include: nuclear surface energy, nuclear surface tension, nuclear forces, nuclear potential well, nuclear pumping, nuclear resonance, nuclear spin, nuclear temperature, nuclear photoeffect, nucleon transfer, and quanta release. All of these examples of quanta transition derive basically from the positive, neutral, and negative potential charges.

When positive and negative energies encounter each other, their energy explodes into quanta release. When positive and neutral charges encounter each other, they attract or conduct as a bridge to another field if it has a negative potential. When neutral and negative charges encounter each other, they attract or conduct as a bridge to another field if it has a positive potential.

Such particles as mesons and gluons—the potential energy caused by the charge transition that has been released or absorbed—are the appropriate quanta energy fields. Though neutral, they carry miniscule positive and negative charges. Since these particles and their fields are so minute, they exhibit a wave state. Once those quanta energy have been released, those baryons will split because the binding force between them is insufficient.

The universal positive and negative charges are not necessarily conservative. Because of its neutral feature, a photon may appear either positive or negative. This variation is caused by variations of the surrounding force field, which generate different types of minute vibrations. It is possible that the charges are non-conservative at an infinitesimal level, yet conservative on a grand scale. If this is so, then conservation of charge is not absolute in the universe; it may vary.

For example, in the 1921 Stern and Gerlach experiment with a specially-designed magnetic pair (one pentagonal and the other rectangular, with a gap between the two poles [the pentagonal point and the rectangular surface]), when a silver-atom beam was passed through the gap, the beam split into two equal, slightly deflected beams.

With regard to that experiment, a possible explanation suggested here is that the excess quanta from the vaporization of heat surrounding the silver atoms caused the quanta groups to be divided equally into positive-inclined and negative-inclined groups through magnetic induced reactions. More specifically, the neutral fields brought into the atoms from the heat tend to compensate the different charges in accordance with the surrounding factorial effect; in this case, the two poles of the magnetic pair. This means the neutral charge could be affected by the different fields, which carry different charges that function to balance the charge; in the above-mentioned case, balancing the charge means splitting the beam to balance not only the outer field charge, but also magnetic polarity.

Some weak forces are interstitial, neutrally-charged fields, which are released from the major binding force, the strong force. In addition, electromagnetic waves are released from atoms that belong to the color charges. In summary, the color force in its entirety and the color charges are based on the positive charge, the neutral charge, and the negative charge.

PART IV

Gravity's Source

Section 1.
Free Fall

EVER SINCE GALILEO'S legendary free-fall experiment from atop the Leaning Tower of Pisa, many similar experiments have been conducted. One such experiment used three differently-sized cannon balls (small, medium, and large).[4] Various other experiments that created a vacuum environment observed the same conclusion. All such free-fall experiments demonstrated that objects all reached the ground at the same time, regardless of differences in weight.

However, when different masses conduct the same movement, the result must differ because the conditions differ. Why, in this experiment, is the observed result always the same? The answer is that the result only appears the same. For example, if 20-kg, 10-kg, and 1-kg cannon balls are taken as free-falling bodies, then (air resistance having always been considered):

$$20 \text{ kg} / (5.98 \times 10^{24} \text{ kg}) \neq 10 \text{ kg} / (5.98 \times 10^{24} \text{ kg}) \neq 1 \text{ kg} / (5.98 \times 10^{24} \text{ kg}).$$

That is, the proportional values between each of the three formulas approach zero; thus, the difference between them is too small to be observed. However, lighter masses should reach the ground's surface faster than heavier ones. (The reason will be explained in Sections 5 and 8.)

Section 2.
The Moon's Face and Volume

That the Moon's same side always faces the Earth implies a 1:1 ratio of its self-rotation and orbit around the Earth. However, one could assume that the Moon once had a faster self-rotation, even after entering the Earth's

---

[4]   This experiment is purported to have been conducted by a Dutch military officer.

orbit. Later on, the Moon could have been struck by meteors, mainly from the shattered planet that once existed between Mars and Jupiter. If some meteors entered into the orbit between the Earth and the Moon, some parts of the Moon's surface obviously could have been impacted by more meteoric masses than the rest. The side most affected would have become heavier, thereby slowing down the Moon's self-rotation. Especially if the meteor entered in between the Earth and the Moon, the self-spin of the Earth's outer circle would have accelerated the meteor toward the moon, while the inner circle of the Moon would have caused the meteor to decelerate and strike the Moon's surface. Thus, the Moon's heavier face is always directed toward the Earth. The heavier side would have caused the Moon to spin incompletely back and forth, slowing down to the point at which the heavier side always faced the Earth.

In this way, the Moon's other side has less surface damage and should be coarser than the heavier side.

With regard to the Moon's volume, certain space missions had recorded surface echoes of up to 15 minutes in length. Therefore, it has been considered that the Moon could be hollow inside.

However, as evidenced by the Earth's ocean tides, the Moon is not hollow; its mass is still about 1/60 that of Earth. The reason for the long duration of the echoes is that the Moon has already been highly developed underneath the crust.

Section 3.
Field of Mass

As discussed in Part III, all levels—extending down from the nucleus to that of the electron—have their fields. Charges from the nucleus permeate every atom through the outer electron layers, which form the surrounding, curved gravity field. Atoms that congregate into molecules or molecules that congregate into mass all have the positive, negative, and neutral charges that originate from the nucleus and penetrate out to form fields of charges. In the fields of the outer layers, the level widths of the three charges, which are extremely minimal, are coseismal membrane layers.

The gravity-field layers consist of layer upon layer of connected positive and negative charged membranes between which are neutral charged fields. The innermost membrane layer is the most condensed, while the outermost is the least condensed. Between each layer is phase contrast. As the superpositioning of layers increases, the charges weaken.

Although the weakest sub-residual charges are positive connected with negative charges, they will not cause "photonic explosion." Yet they can join tightly with tensile elasticity and tensility. This field neither oscillates nor flutters since it has already become matter's outermost final state.

For example, the rarefied zone of the Earth's gravitational field is partially incomplete. The outermost edge will change minutely depending on changes in energy and matter (e.g., a meteor strike will add to the Earth's volume, which will increase the gravity field's range or the combined gravity fields of the Earth and the Moon will extend farther into space than the Earth alone could do).

Section 4.
Inertia

All matter has its outer field, which corresponds to its proportional level value. That value is directly connected to the total charges from the matter's atoms, which form its sub-residual field. Thus, layer upon layer, the field extends to its limit, determined by the volume of its mass. This can be called that matter's own gravity field; that is, its inertia field.

When an object accelerates in a linear direction, its inertia field (its own gravity field) has its vector direction, which causes its field's layers to compress; simultaneously, the counter-direction of its field's layers expands. The linear motion of inertia can be pictured thusly: Imagine a ball at rest. The concentric membranes of its inertia field are round. Next, imagine that the ball is pushed; the inertia field's membranes become ellipsoid in shape. When the ball is in accelerated motion, it can be compared to a boiled egg. Like the yoke, the ball is still round; like the egg white, the inertia field is ellipsoid (except that the egg white has only one membrane layer while the inertia field has many) (Fig. h). The forward-moving side of the field is more condensed, while the dragging side is more elongated or loosened. Although the shape of the compressed and opposite extending sides changes minutely, between each layer, the combined repulsive force equals the accelerating matter's G-force. Whenever the acceleration ceases, the surrounding inertia field will resume its original state at the speed of light and maintain that speed moving forward in a straight line (Fig. i). Each time an object accelerates, the G-force of its surrounding inertia layers will resist acceleration. That is, when an object is at rest, it will remain still; when it is moving, it will continue moving linearly. With regard to an object moving in curved or circular direction, the object's surrounding inertia field will turn in hypocycloid motion and continue

Fig. h

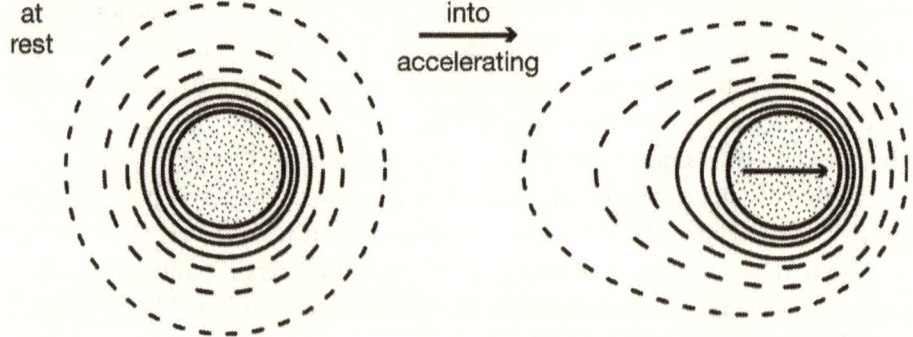

Fig. i

linear motion at
a constant speed

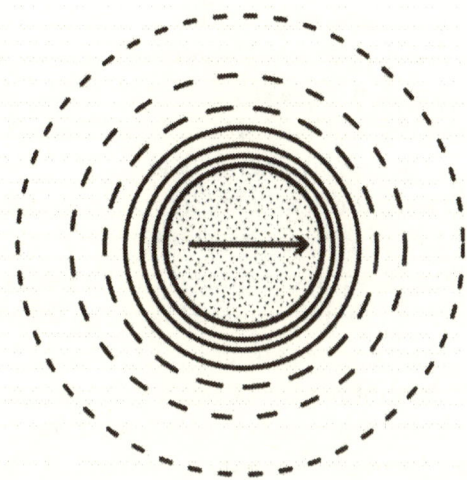

changing direction; that is, the vectors of the membrane layers' angular momentum will continue changing along with the object. Whenever the moving or turning force ceases, the object will follow the compressed side and continue moving linearly, tangential to the curved movement (at a right angle to the center); this exemplifies centrifugal force.

If an object travels slower than the speed of light, its inertia field can resume its original shape when the object is at rest. If an object accelerates instantaneously at the speed of light, its inertia field will be unable to resume its original state; however, the inertia field's charges will press the object directly, while the other side of the inertia field will pull off and disintegrate the structure of the object's atoms. This force is strong enough to dissemble the structure of the object's atoms into "photonic explosion," radiating in all directions. Therefore, if a spaceship were to accelerate at the speed of light immediately, not only would time not stop; the spaceship and its travelers would disintegrate into "photonic" quanta, including neutrinos, etc.

However, if a spaceship were to accelerate slower than the speed of light continuously, not only would it not disintegrate when it reached the speed of light—because its inertia field would have had enough time to recover its original state at the speed of light; it would exceed the speed of light without limit. Moreover, time would not stop; it would slow.

When Newton considered inertia itself equal to the gravitational force, he was not totally wrong. Likewise, Einstein was not wrong when he thought there were substances in the center of the inertia force resistant to the acceleration tendency because all matter have their fields, the source of which is the field's center, which are atomic charges. At the turn of the last century, Poincare thought that inertia or the resistance to accelerate is not caused completely by electromagnetic energy, but by all energy features; he was totally correct.

Section 5.
Gravity Force

All matter has its respective surrounding field; that is, gravity field or inertia field, which has multiple membrane layers. In addition, each layer has its "sub-charges," which exert contracting and repulsive forces. Therefore, in Earth's gravity field, any free-falling body's inertia field is affected by the Earth's enormous, condensed layers of compression. Although the inner layers of the Earth's repulsive force counteract the falling body's lower inertia field, the effect of this force is not as great in dimension as that of the outer circle of the Earth's G-field onto the falling body's upper inertia field. Thus, the body falls to Earth (Fig. j).

As long as the falling body's mass is greater, its surrounding inertia field will also be greater. Because of the larger dimension of the falling body's lower inertia field, which encounters the Earth's repulsive gravity layer, the falling speed will be impeded and thus slower. If the falling body's mass equals that of Earth, then compression and repulsive forces will be equivalent. In this case, the falling body and Earth will glide around each other. For example, even though the Moon's mass is 1/60 that of Earth, the Moon glides around the Earth, rather than falling to Earth; only, the Moon's gliding around the Earth is greater than the Earth's gliding around the Moon. This explanation differs slightly from Newton's reasoning that the Moon keeps falling to Earth, while balanced by the centrifugal force.

Although Galileo concluded that all free-falling bodies (no matter the volume) reach the ground at the same time, they only appear to do so (see Section 1). However, if a heavier body were crushed into small particles, then all of those particles would become individual, light falling bodies that would reach the ground faster than their original heavy one. (The reason will be discussed in Section 6.) If all of the loose particles were gathered into one pack and then dropped from the same height, the speed would be slower than that of the individual particles. If all of the particles were tightly compacted into one object, then the falling speed would be even slower. Furthermore, if they were compacted into one matter smaller than an atom, then that falling matter and its surrounding inertia field would become so large that even the Earth could possibly fall into it. The reason is that the falling matter's inertia or gravity field is directly touched by its most powerful nuclear binding force, which far exceeds Earth's gravity force.

All extra-terrestrial bodies have their inertia or gravity fields; though their functions are more complex, the underlying principles are the same as those described above.

Section 6.
Gravity Field of the Solar System

The Moon's Earth-facing side has more mass than its opposite side; its surrounding inertia or gravity field of the membrane layers is denser than that of the opposite side and its repulsive force is stronger. In addition, it can slide up against and around the membranes of the Earth's gravity field. The back of its surrounding field can merge with the outer membrane of

Fig. j

Much merging with
Earth's layers and
contracting, but
continuing to repel
downward and changing
into new layers

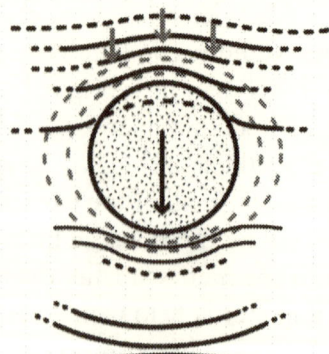

Less merging and
repelling occur
because of opposite
curvature.

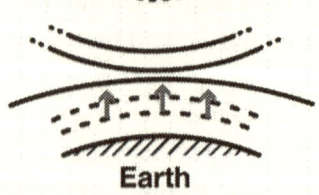

Earth

Earth's gravity field. And together, as a whole terrestrial body, they orbit the Sun.

Because of the Sun's gaseous structure, all of its atoms and molecules have ever-changing vectors. All of the particles in the surrounding field cannot easily merge into one body. The Sun's surface inertia field, like that of the crushed, free-falling particles mentioned in Section 5, is not strong. However, the Sun has enormous mass. For this reason, its outer membrane layers can become connected, enforcing the gravity force to the outermost layers. These membrane layers become increasingly more organized and more connected until the membrane is totally connected; it then becomes denser. Thus, the Sun's farthermost inertia or gravity field is strong. The planet Mercury, positioned at the edge of the Sun's strong gravity field, is squeezed by it into a streamline design, which causes the planet to orbit the Sun more quickly (Fig. k).

As mention in Section 4, every matter has its inertia field. Whenever an object moves, its surrounding inertia field will change shape to follow. When any star or planet moves, its inertia field will change since the surrounding inertia field is itself a gravity field. Moreover, that field is irrotational because it derives from the field source, matter, and its atoms, constructed by their positive, neutral, and negative charges. Therefore, the field will follow its original atoms. However, when the terrestrial body moves, the field will follow its movement; if the body turns or spins, the field will become part of the terrestrial body's rotational field.

When a terrestrial body, such as a planet, is at rest, its surrounding gravity field membrane layers have their contracting force to maintain the spherical shape of the gravity field. In this state, any satellite or rock will remain scattered, according to its orbiting speed. However, when the planet spins, centrifugal force will be added to the gravity field. That motion will make the gravity field spin out toward the equator plane, while the top and bottom close to the imaginary axle will be pulled down toward the planet's surface. Thus, the membranes closer to the imaginary axle are denser, while those closer to the equator are looser, with more space between the membrane layers. Thus, the gravity field becomes ellipsoid in shape. Any satellite or rock will be pushed down toward the loosened equatorial plane (Fig. l). This explains, for example, why Saturn's equatorial plane has so many layers of rings with orbiting objects.

However, a satellite with a heavy mass on the equator's plane will have its own stronger gravity field, and its orbit around the planet will be elliptical.

Fig. k

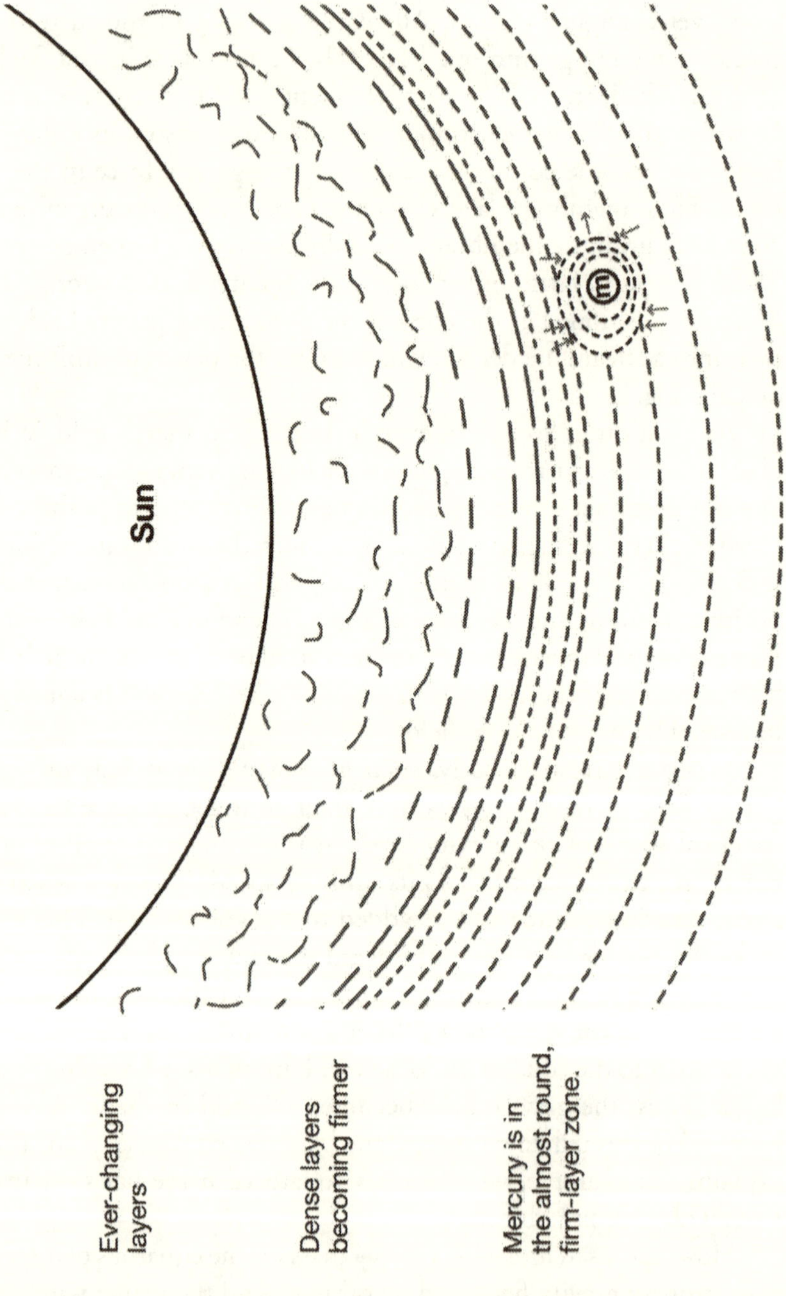

Sun

Ever-changing layers

Dense layers becoming firmer

Mercury is in the almost round, firm-layer zone.

Fig. I

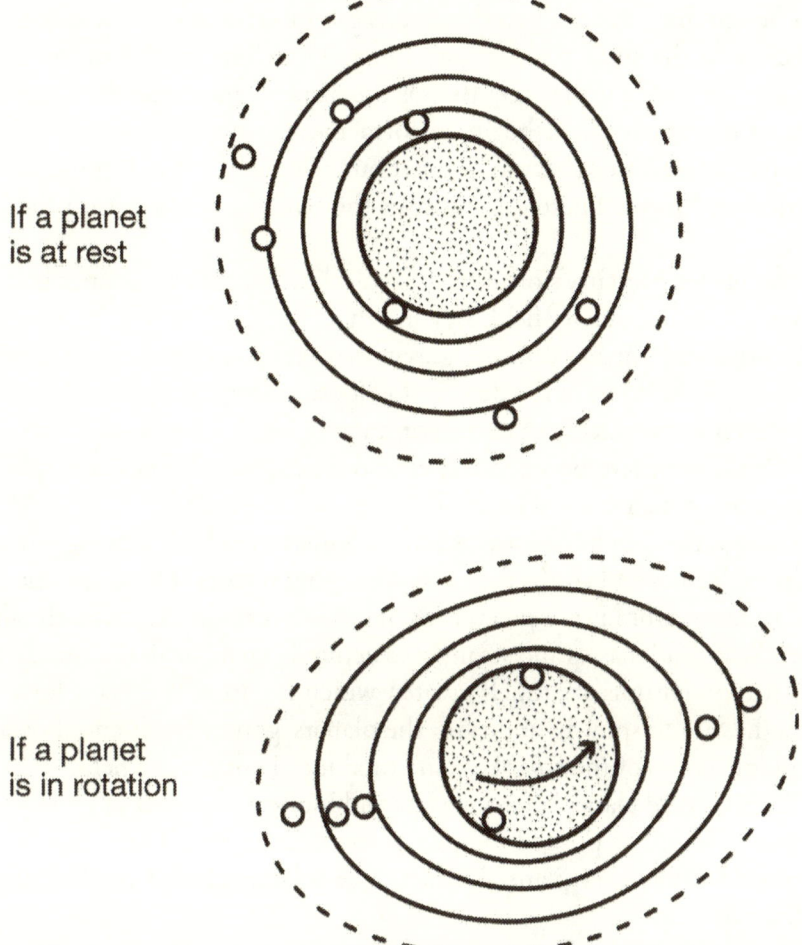

If a planet
is at rest

If a planet
is in rotation

In our solar system, all the planets and their many satellites have their own gravity fields; therefore, the total gravity field of the solar system is ever-changing. This is why Einstein thought that space is curved. However, space does not curve. But, according to the gravity field's ever-changing status, the gravity field will curve. Space and gravity are different concepts.

The structure of the gravity field is smaller than neutrinos, even though it still has material movement. Therefore, neutrinos can be called matter. It is only that the neutral neutrinos pass through the gravity field as if nothing were there. Yet, gravity field is the smallest matter in this universe.

Though the Sun has a grand mass, its gaseous structure causes the gravity field close to the surface to be less strong. Therefore, its flames can spread out quite a distance into space. But, as the distance increases, the membrane layers of the gravity field begin to form more strongly, and those gravity layers just formed are spherical. Farther out, they tend to be ellipsoid because of centrifugal force. Therefore, most of the Sun's planets orbit on the same plane.

The planet Mercury's orbit has a 7° decline because the planet is still in the Sun's spherical gravity field. Beyond that, The Sun's gravity field tends to be ellipsoid. Still farther out, the gravity field tends to revert to be spherical because, in those layers of less density and loosened structure, centrifugal force will weaken more quickly than the contracting force. The planet Pluto orbits in a zone somewhere between the ellipsoid field membranes and spherical membranes, which cause its orbit to have a 17.1° decline.

Though the Sun has a great mass, it consists mainly of moving particles; and its molecules and atoms have ever-changing vectors. Therefore, the Sun's angular momentum is not great; thus, it cannot turn fast. Conversely, all the planets have solid masses and strongly structured gravity fields. Naturally, their angular momentums are much greater, which is part of the main force that causes the Sun to spin. In addition, the planets' gravity fields could occur in the middle layers of the solar system's total inertia or gravity field.

The shattered planet between Mars and Jupiter lost its major gravity field. Eventually, it will be pushed to other planets or the Sun. Thus, its orbit will be embedded by other planets' orbits. The solar system planets' orbits will be affected.

Comets are affected by the Sun's gravity force and planets' gravity force; therefore, a comet's orbit will exhibit changes. A comet travels from outer spherical through ellipsoid and into inner spherical of the solar system's gravity field; it then accelerates in the inner spherical and projects out through the ellipsoid back into the greater spherical field.

Section 7.
Galaxies and Beyond

Because the gravity fields of all terrestrial masses turn, they expand; as gravity fields expand, they simultaneously push neighboring fields, which are still expanding. The superpositioned repulsive forces cause the outer galaxies to expand at a speed that exceeds the speed of light.

The volume of extraterrestrial masses is gigantic, and the distances between the masses are too large for their gravity fields to congregate. Thus, their repulsive forces are greater than their contracting forces, which explain why, so far, the universe is still expanding. However, at the extremity, there could possibly exist extremely weak, congregated gravity membranes of the whole universe that could constrain those expanding masses.

Section 8.
Miscellaneous Conclusions

Free fall is caused by the G-force. The smaller the mass of the free-falling body, the faster it accelerates; conversely, the larger its mass, the slower it accelerates. If the volume of the mass is equal to that of the planet, then the free-falling body and the planet will revolve around each other circularly in the same plane. If their volumes differ, the smaller mass will revolve elliptically around the larger one, while the larger one will also be affected by the smaller one. For example, although the Earth does not revolve around the Moon, it is still affected slightly by the Moon's gravity field and surrounding angular momentum, and makes a small orbit. Three is the maximum number of masses that can revolve around each other in the same plane. Because the entire universe is either spinning or turning, it tends to exist in a plane, rather than a sphere.

The gravity field is not a gravitational attraction field; rather, it is an inertia repulsive force field. When any free-falling body accelerates toward the Earth, the top membrane layers of the free-falling body are pushed by the Earth's membrane layers; their successive membrane layers repel each other downward, while the lower membrane layers of the free-falling body have the opposite curvature against the Earth's curvature (Fig. j). However, because the top curvature dimensions are larger than the bottom ones, the top layers' repulsive force is greater than that of the bottom layers, which causes the body to fall to Earth. The gravity field is repulsive force field; thus, all objects—large or small—have resistance to compressive strain.

The Moon's orbiting of the Earth is maintained by three forces: 1) the mutual outer membrane layers of the inertia fields, which hold each other (repulsive force); 2) the inner membrane layers' repelling of each other, which causes the Moon to glide around the Earth (repulsive force); and 3) the Moon's centrifugal force, which adds to the mutual repulsive force, causing the Moon to be projected away from the Earth (inertia force). Together, these three forces maintain the Moon's orbiting of the Earth. Thus, even though the Moon may be affected by other outside forces (e.g., passing by other planets' gravity fields or meteors added to the Moon or Earth), it will not be easily pushed off track or fall to Earth. If the Moon's orbit around the Earth were caused only by the Moon's centrifugal force and the Earth's attraction force, then many terrestrial bodies would be out of balance and would stop orbiting each other because of other gravity fields' interference or meteors added to the mass of planets. That is, the Moon's centrifugal force and the Earth's attraction force are too fragile to maintain balance.

All masses have their gravity fields. The dense, extremely thin membranes of a mass's surface possess homogeneous positive and negative charges; and between the membranes, extremely minute neutral fields insulate them superpositionally, layer upon layer, to form the mass's largest field. Altogether, they can be called total mass. Because the positive, negative, and neutral charges are extremely small, they cannot annihilate each other into light or initiate other stronger functions. Since these three charges are permeated by atoms, the outer gravity field becomes a part of the three charges' matter; only, this part, gravity field, is invisible. One cannot eliminate the field without eliminating its source, which is matter.

If one would wish to isolate the Earth's field, for example, one could possibly use a purified light field, whose photons are in a self-congruent state or use an even more strongly charged, condensed outer field, such as a fast-spinning one, to create a strongly expanding gravity field. On Earth's surface, the attraction force from the center of the Earth is not actually so; just as Newton said, there is no gravity in the Earth's center. However, one can still determine the weight of an object on the Earth's surface by using an imaginary diameter that passes through the Earth's center as the route with which to calculate the Earth's mass, minus the centrifugal force of the Earth's motion. This may sound too complicated to do; fortunately, scales have already accounted for these factors automatically. Because the Earth's mass density differs by zone, different field intensities will form curvature differences.

The inertia field and gravity field are not different; therefore, one can say that the inertia or gravity field is a part of a mass. While that part is much weaker, its tenacity is extremely greater. Therefore, affecting this field will directly affect the matter itself. For example, a pendulum in a high-voltage field will swing slower because the pendulum's inertia field is affected by the high-voltage's electrical charges field, which is much stronger than the pendulum's charges field; that is, its inertia field. It is not necessary for the high-voltage field to physically push the pendulum in order to affect it.

Part II, Section 3, which discusses how a photon travels, mentions that light will curve when passing a terrestrial body's gravity field. The reason is that a photon has equal positive and negative charges and carries its own wave field, which is neutral. This neutral wave field is actually the photon's own gravity field. When a photon travels through a terrestrial body's outer gravity field, which consists of sub-neutral spherical layers, the photon's neutral charge will be affected by the neutral membrane of those gravity layers and will be repelled along the curvature. Moreover, the photon's carrying wave field is repelled by the gravity membrane layers' neutral insulation. Thus, a photon travels a curved path.

The gravity field still has the feature of matter. Whenever a terrestrial body turns, its gravity field will likewise follow and turn; it will also expand laterally. Similarly, if the gravity field could be stopped, then the turning of the terrestrial body would also stop. When a terrestrial body turns, its gravity field will usually be flattened (like an M&M chocolate). Most solar systems look like that. However, different systems can affect each other, causing their gravity fields to become misshapen into multi-curved fields. Therefore, terrestrial bodies' gravity fields can change to curve irregularly. Thus, the gravity field can be curved; however, space cannot be curved. It is said that the shortest travel distance between two points is a geodetic line; this statement is not always true. When one travels out of the gravity field, the line of travel is still straight.

The gravity field's repulsive force can be multiplied to reinforce and accelerate the expanding speed of the terrestrial system.

When an electron leaves an atom, it will carry some of the neutral charge field from the atom; therefore, its mass will be heavier than the electron in a state of rest. Likewise, when the gravity field turns in accordance with its planet turning, the gravity field itself will be heavier because it absorbs some of the planet's energy field.

Parts III and IV have both referred to the three charges: positive, neutral, and negative. The law of parity deals only with positive and negative charges;

however, the neutral charge also pertains to all existence. For example, a neutral charge sometimes can remain with the already balanced positive and negative charges. At other times, it can interfere with either positive or negative charges. Therefore, the law of uncertainty still holds true.

On the Earth's surface, all matter and its inertia field are already combined with the membrane layers of the Earth's gravity field. Therefore, that matter is minimally affected by the Earth's movements in space (including rotation and orbiting). However, once the matter accelerates, its own inertia field will split from the membrane layers of the Earth's gravity field. Therefore, it resists acceleration.

Section 9.
Conclusion

The gravity field consists of positive, neutral, and negative charges in a subresidual, total combined field. This is the lowest level of the three-charges field. It is in a quasi-neutral state, which is the weakest level of all matter. Even though this level is the weakest, it still begins from the most saturated membrane layer and, layer upon layer, develops outward. The innermost layers have greater density, which gradually decreases in the outer layers.

The Earth's magnetic field consists of positive, neutral, and negative charges as well. However, it is a linear field, rather than a spherical membrane field; therefore, it has less interaction with the gravity field since the gravity field is a membrane field.

Part I, Section 12, has already discussed the unlikelihood of Earth changing polarities periodically; however, it mentions that recorded observations have revealed changes in magnetic polarities, for which there is no explanation. If the Sun's magnetic storms have struck the Earth multiple times—the Sun does change its magnetic polarities—then those storms would be large enough to cover the entire area surrounding the Earth. In this case, the storms would not cause any biased strike—that is, they would not hit one side, which might cause the polarities of the Earth's magnetic field to change. Therefore, the Earth's magnetic polarities will most likely remain unchanged.

With regard to recorded observations of changes in the polarities of Earth's magnetic field, the explanation provided here is as follows: When the lava split from deep ditches under the ocean or on the surface of the Earth's crust, the lava from both sides cooled nearly evenly. Once cooled, the magnetic metals in the rocks were induced by the older magnetic rocks underneath or alongside. This natural process is analogous to what happens when a magnetic bar is touched by a simple iron bar, which has no magnetic force in it. Once it touches the magnetic bar, the iron bar will be induced and magnetized with opposite polarities of the original magnetic bar. Likewise, each new lava flow that cools will always be induced to the opposite polarity to the previous induced and magnetized rock. Therefore, the polarity change is not periodic per se. However, as long as lava continues to push out through the ditches, a polarity change will always occur when the cooled lava becomes rock through induction from the older rock.

Any matter has its own gravity field. The smaller the matter, the smaller the curvature of its gravity field. The denser the matter, the tighter its gravity field and the more membrane layers it will have. For any irregularly-shaped matter, the membrane layers close to its surface may be irregularly shaped; however, the outward membrane layers will tend to be round and spherical.

For any flowing matter (e.g., fluid, air, or free particles), the immediately surrounding layers of the gravity field are not the strongest ones because their vectors conflict. However, the membrane layers, moving from inner to outer, will gradually reach a first-saturated layer; that is the strongest layer. From that layer outward, the tendency of the layers is a gradual weakening.

Each orbiting terrestrial body is in its least pressured route to glide in other gravity fields.

Although the gravity field can expand or contract according to its center matter's movement, the gravity field layers could also be isolated by an even stronger field of light, electricity, gravity, or other tightly connected fields.